tredition®

tredition was established in 2006 by Sandra Latusseck and Soenke Schulz. Based in Hamburg, Germany, tredition offers publishing solutions to authors and publishing houses, combined with worldwide distribution of printed and digital book content. tredition is uniquely positioned to enable authors and publishing houses to create books on their own terms and without conventional manufacturing risks.

For more information please visit: www.tredition.com

TREDITION CLASSICS

This book is part of the TREDITION CLASSICS series. The creators of this series are united by passion for literature and driven by the intention of making all public domain books available in printed format again - worldwide. Most TREDITION CLASSICS titles have been out of print and off the bookstore shelves for decades. At tredition we believe that a great book never goes out of style and that its value is eternal. Several mostly non-profit literature projects provide content to tredition. To support their good work, tredition donates a portion of the proceeds from each sold copy. As a reader of a TREDITION CLASSICS book, you support our mission to save many of the amazing works of world literature from oblivion. See all available books at www.tredition.com.

 Project Gutenberg

The content for this book has been graciously provided by Project Gutenberg. Project Gutenberg is a non-profit organization founded by Michael Hart in 1971 at the University of Illinois. The mission of Project Gutenberg is simple: To encourage the creation and distribution of eBooks. Project Gutenberg is the first and largest collection of public domain eBooks.

The American Type of Isthmian Canal Speech by Hon. John Fairfield Dryden in the Senate of the United States, June 14, 1906

John F. (John Fairfield) Dryden

Imprint

This book is part of TREDITION CLASSICS

Author: John F. (John Fairfield) Dryden
Cover design: Buchgut, Berlin – Germany

Publisher: tredition GmbH, Hamburg - Germany
ISBN: 978-3-8472-1275-1

www.tredition.com
www.tredition.de

Copyright:
The content of this book is sourced from the public domain.

The intention of the TREDITION CLASSICS series is to make world literature in the public domain available in printed format. Literary enthusiasts and organizations, such as Project Gutenberg, worldwide have scanned and digitally edited the original texts. tredition has subsequently formatted and redesigned the content into a modern reading layout. Therefore, we cannot guarantee the exact reproduction of the original format of a particular historic edition. Please also note that no modifications have been made to the spelling, therefore it may differ from the orthography used today.

THE AMERICAN TYPE OF ISTHMIAN CANAL

HON. JOHN FAIRFIELD DRYDEN

THE JOHN F. DRYDEN STATUE

The above is a picture of the bronze statue of the late United States Senator John F. Dryden, Founder of The Prudential and Pioneer of Industrial insurance in America, erected by the John F. Dryden Memorial Association, with this inscription: *"A tribute of esteem and affection from the field and office force."* The statue is located at the Home Office of The Prudential, Newark, N.J., and is unique, being the gift of a staff of over 16,000 employees. It cost $15,000. The sculptor was Karl Bitter.

No. 8

PANAMA-PACIFIC EXPOSITION
MEMORIAL PUBLICATIONS OF THE PRUDENTIAL
INSURANCE COMPANY OF AMERICA

THE AMERICAN TYPE
OF
ISTHMIAN CANAL

SPEECH BY
HON. JOHN FAIRFIELD DRYDEN
IN THE SENATE OF THE UNITED STATES
JUNE 14, 1906

1915
PRUDENTIAL PRESS, NEWARK, NEW JERSEY

The ancient "Dream of Navigators" has at last been realized in the completion and successful operation of the PANAMA CANAL, fittingly commemorated by the Panama-Pacific International Exposition. Among the men who contributed in a measurable degree to the attainment of this national ideal was the late United States Senator, John F. Dryden, *President of THE PRUDENTIAL. As a member of the Senate Committee on Interoceanic Canals, Mr. Dryden, after mature and extended consideration, gave the weight of his influence and vote in favor of the lock-level principle of canal construction. The lock-level type was finally decided upon, although the majority of Mr. Dryden's conferees and the International Board of Consulting Engineers at first strongly favored the sea-level type. By his determined support of the one and his well-reasoned opposition to the other, Mr. Dryden was able to secure the enactment of legislation in accordance with his views and to bring about the completion of this tremendous undertaking within our time, thus leaving a permanent imprint upon the country's history.*

THE AMERICAN TYPE OF ISTHMIAN CANAL

It was on June 14, 1906, when the Canal subject was up for final consideration, that Mr. Dryden addressed the Senate. The official records show that "S. 6191, to provide for the construction of a sea-level canal connecting the waters of the Atlantic and Pacific oceans, and the method of construction," was before Congress, and it was in opposition to this measure that Mr. Dryden patriotically pledged his devotion to American enterprise and American ability by declaring for the lock-level type of canal, built by American engineers and under American supervision, concluding with the following words, which deserve to be recalled on this memorable occasion as a tribute to the native genius and enterprise of the American people:

"I am entirely convinced that the judgment and experience of American engineers in favor of a lock canal may be relied upon with entire confidence and that such an enterprise will be brought to a successful termination. I believe that in a national undertaking of this kind, fraught with the gravest possible political and commercial consequences, only the judgment of our own people should govern, for the protection of our own interests, which are primarily at stake. I also prefer to accept the view and convictions of the members of the Isthmian Commission, and of its chief engineer, a man of extraordinary ability and large experience. It is a subject upon which opinions will differ and upon which honest convictions may be widely at variance, but in a question of such surpassing importance to the nation, I, for one, shall side with those who take the American point of view, place their reliance upon American experience, and show their faith in American engineers."

The Panama Canal problem has reached a stage where a decision should be made to fix permanently the type of the waterway, whether it shall be a sea-level or a lock canal. An immense amount of evidence on the subject has in the past and during recent years been presented to Congress. An overwhelming amount of expert opinion has been collected, and an International Board of Consulting Engineers has made a final report to the President, in which experts of the highest standing divide upon the question. The Senate Committee on Interoceanic Canals has likewise divided. It is an issue of transcendent importance, involving the expenditure of an enormous sum of money, and political and commercial consequences of the greatest magnitude, not only to the American people, but to the world at large.

The report of the International Board has been printed and placed before Congress. A critical discussion of the facts and opinion presented by this Board, all more or less of a technical and involved nature, would unduly impose upon the time of the Senate at this late day of the session. In addition, there is the testimony of witnesses called before the Senate committee, which has also been printed in three large volumes, exceeding 3,000 pages of printed matter. To properly separate the evidence for and against one type of canal or the other, to argue upon the facts, which present the greatest conflict of engineering opinion of modern times, would be a mere waste of effort and time, since the evidence and opinions are as far apart and as irreconcilable as the final conclusions themselves. It is, therefore, rather a question which the practical experience and judgment of members of Congress must decide, and I have entire confidence that the will of the nation, as expressed in its final mandate, will be carried into successful execution, whether that mandate be for lock canal or sea-level waterway.

The Panama Canal presents at once the most interesting and the most stupendous project of mankind to overcome by human ingenuity "what Nature herself seems to have attempted, but in vain." From the time when the first Spanish navigators extended their explorations into every bay and inlet of the Central American isthmus, to discover, if possible, a short route to the Indies, or "from Cadiz to Cathay," the human mind has not been willing to rest content and accept as insurmountable the natural obstacles on the Isthmus which prevent uninterrupted communication between the Atlantic and the Pacific. Excepting, possibly, Arctic explorations, in all the romantic history of ancient and modern commerce, in all the annals of the early navigators and explorers, there is no chapter that equals in interest the never-ceasing efforts to make the Central American isthmus a natural highway for the world's commerce—a direct route of trade and transportation from the uttermost East to the uttermost West.

As early as 1536 Charles V ordered an exploration of the Chagres River to learn whether a ship canal could not be substituted for an existing wagon road, and Philip II, in 1561, had a similar survey made in Nicaragua for the same purpose. From that day to this the greatest minds in commerce and engineering have given their attention to the problem of an interoceanic waterway; every conceivable plan has been considered, every possible road has been explored, and every mile of land and sea has been gone over to find the best and most practical solution of the problem.

The history of these early attempts is most interesting, but it is no longer of practical value, for it has no direct bearing upon present-day problems. Most of the efforts were wasted, and many of them were ill advised, but the present can profitably consider the more important lessons of the past. It was written in the book of fate that this enterprise, the most important in the world of commerce and navigation, should be American in its ending as it had been in its practical beginning. From the day when the first train of cars crossed the Isthmus from Panama to Aspinwall, to facilitate the transportation of passengers and freight across the narrow belt of land connecting the northern and southern continents, the imperative necessity of a ship canal was made apparent. Just as the railway followed the earlier wagon roads of the Spanish adventurers, so a ship canal will naturally succeed or supplement the railway.

Natural conditions on the Isthmus materially enhance the physical difficulties to be overcome in canal construction. Even the precise locality or section best adapted to the purpose has for many years been a question of serious doubt. The Isthmus of Tehuantepec, the Nicaraguan route, the utilizing of a lake of large extent, and finally the narrow band of land and mountain chain at Panama, each offers distinct advantages peculiar to itself, with corresponding disadvantages or local difficulties not met with in the others. Many other projects have been advanced; in all, at least some twenty distinct routes have been laid out by scientific surveys, but the most eminent American engineering talent, considering impartially the natural advantages and local obstacles of each, finally, in 1849, decided upon the isthmus between the Bay of Panama and Limon Bay as the most feasible for the building of the railroad, and some fifty years later for the building of the Isthmian Canal. Every further study, survey, and inquiry has confirmed the wisdom of the earlier choice, which has been adopted as the best and as the permanent plan of the American government, which is now to build a canal at the expense of the nation, but for the ultimate benefit of all mankind.

The Panama railway marked the beginning of a new era in the history of interoceanic communication. The great practical usefulness of the road soon made the construction of a canal a commercial necessity. The eyes of all the world were upon the Isthmus, but no nation made the subject a matter of more profound study and inquiry than the United States. One surveying party followed another, and every promising project received careful consideration. The conflicting evidence, the great engineering difficulties, the natural obstacles, and, most of all, the Civil War, delayed active efforts; but public interest was maintained and the general public continued to view the project with favor and to demand an American canal.

During the seventies a French commission made surveys and investigations on the Isthmus which terminated in the efforts of De Lesseps, who undertook to construct a canal, and, in 1879, called an international scientific congress to consider the project in all its aspects and determine upon a practical solution. The United States was invited to be represented by two official delegates, and accordingly President Hayes appointed Admiral Ammen and A.C. Menocal, of the United States Navy, both of whom had been connected with surveys and explorations on the Isthmus. Mr. Menocal presented his plan for a canal by way of Nicaragua, but it was evident that the Wyse project, of a canal by way of the Isthmus of Panama, had the majority in its favor, and the only question to determine was whether the canal to be constructed should be a sea-level or a lock canal. The American delegates were convinced, in the light of their knowledge and experience, that a sea-level canal would be impracticable, if not impossible. In this they were seconded by Sir John Hawkshaw, a man thoroughly familiar with canal problems, and who exposed the hopelessness of an attempt to make a sea-level ship canal, pointing out that there would be a cataract of the Chagres River at Matachin of 42 feet, which in periods of floods would be 78 feet high, and a body of water that would be 36 feet deep, with a width of 1,500 feet.

Opposition to the sea-level project proved of no avail. The facts were ignored or treated with indifference by the French, who were determined upon a canal at Panama and at sea level, resting their conclusions upon the success at Suez, with which enterprise many of those present at the congress, in addition to De Lesseps, had been connected. But the problems and conditions to be met on the Isthmus of Panama were decidedly different from those at Suez, and subsequent experience proved the serious error of the sea-level plan as finally adopted. The congress included a large assemblage of non-professional men, and of the French engineers present only one or two had ever been on the Isthmus. The final vote was seventy-five in favor of and eight opposed to a sea-level canal. Rear-admiral Ammen said: "I abstained from voting on the ground that only able engineers can form an opinion *after careful study* of what is actually possible and what is relatively economical in the construction of a ship canal." Of those in favor of a sea-level canal not one had made a practical and exhaustive study of the facts. The project at this stage was in a state of hopeless confusion. In spite of these obstacles, De Lesseps, with undaunted courage, proceeded to organize a company for the construction of a sea-level canal.

As soon as possible after the adjournment of the scientific congress of 1879 the Panama Canal Company was organized, with Ferdinand De Lesseps as president. The company purchased the Wyse concession, and by 1880 sufficient funds had been secured to proceed with the preliminary work. The next two years were used for scientific investigations, surveys, etc., and the actual work commenced in 1883. The plan adopted was for a sea-level canal having a depth of 29.5 feet and a bottom width of 72 feet. This plan in outline and intent was adhered to practically to the cessation of operations in 1888.

In that year operations on the Isthmus came to an end for want of funds. The failure of the company proved disastrous to a very large number of shareholders, mostly French peasants of small means, and for a time the project of interoceanic communication by way of Panama seemed hopeless. The experience, however, proved clearly the utter impossibility of private enterprise carrying forward a project of such magnitude and which had attained a stage where large additional funds were needed to make good enormous losses, due to errors in plans, to miscarriage of effort, and, last but not least, to fraud on stupendous scale. With admirable courage, however, the affairs of the first Panama Canal Company were reorganized, after the appointment of a receiver, on February 4, 1889. A scientific commission of inquiry was appointed to reinvestigate the entire project and report upon the work actually accomplished and its value in future operations. The commission, made up of eminent engineers, sent five of its members to the Isthmus to study the technical aspects of the problem, and a final report was rendered on May 5, 1890. The recommendation of the commission was for the construction of a canal with locks, the abandonment of the sea-level idea, and for a further and still more thorough inquiry into the facts, upon the ground that the accumulated data were "far from possessing the precision essential to a definite project." This took the project of canal construction out of the domain of preconceived ideas based upon guesswork into the substantial field of a scientific undertaking for commercial purposes. The receiver at once commenced to reorganize the affairs of the company, and accordingly, on October 21, 1894, the new Panama Canal Company came into existence under the general laws of France. The charter of the new company provided for the appointment of a technical committee to formulate a final project for the completion of the canal. This committee was organized in February, 1896, and reached a unanimous conclusion on November 16, 1898, embodied in an elaborate report, which is probably the most authoritative document ever presented on an engineering subject. The recommendation of the commission was unanimously in favor of a lock canal. [1]

The subsequent history of the De Lesseps project and the American effort for a practical route across the Isthmus are still fresh in our minds and need not be restated. The Spanish-American war and the voyage of the *Oregon* by way of Cape Horn, more than any other causes, combined to direct the attention of the American people to conditions on the Isthmus, and led to the public demand that by one route or another an American waterway be constructed within a reasonable period of time and at a reasonable cost. It will serve no practical purpose to recite the subsequent facts and the chain of events which led to the passage of the act of March 3, 1899, which authorized the President to have a full and complete investigation made of the entire subject of Isthmian canals.

A million dollars was appropriated for the expenses of a commission, and in pursuance of the provisions of the act the President appointed a commission consisting of Rear-admiral Walker, United States Navy, president, and nine members eminent in their respective professions as experts or engineers. A report was rendered under the date of November 30, 1901, in which the cost of constructing a canal by way of Nicaragua was estimated at $189,864,062 and by way of Panama at $184,233,358, including in the last estimate $40,000,000 for the estimated value of the rights and property of the New Panama Canal Company. The company, however, held its property at a much higher value, or some $109,000,000, which the Commission considered exorbitant, and thus the only alternative was to recommend the construction of a canal by way of the Nicaraguan route. Convinced, however, that the American people were in earnest, the New Panama Company expressed a willingness to reconsider the matter, and finally agreed to the purchase price fixed by the Isthmian Commission.

By the Spooner act, passed June 28, 1902, Congress authorized the President to purchase the property of the New Panama Canal Company for a price not exceeding $40,000,000, the title to the property having been fully investigated and found valid. The Isthmian Commission, therefore, recommended to Congress the purchase of the property, but the majority of the Senate Committee on Interoceanic Canals disagreed, and it is only to the courage and rare ability of the late Senator Hanna and his associates, as minority members of the committee, that the nation owes the abandonment of the Nicaraguan project, the acquirement of the Panama Canal rights at a reasonable price and the making of the project a national enterprise.

The report of the minority members of the Senate committee was made under date of May 31, 1902. It is, without question, a most able and comprehensive dissertation upon the subject, and forms a most valuable addition to the truly voluminous literature of Isthmian canal construction. The report was signed by Senators Hanna, Pritchard, Millard, and Kittredge. "We consider," said the committee, "that the Panama route is the best route for an isthmian canal to be owned, constructed, controlled, and protected by the United States." It was a bold challenge of the conclusions of the majority members of the committee, but in entire harmony with and in strict conformity to the views and final conclusions of the Isthmian Commission. The minority report was accepted by the Congress and a canal at Panama became an American enterprise for the benefit of the American people and the world at large.

Such, in broad outline, is the present status of the Panama Canal. A grave question presents itself at this time, which demands to be disposed of by Congress, and to which all others are subservient. Shall the waterway be a sea-level or a lock canal? It is a question of tremendous importance—a question of choice equally as important as the one of the route itself. A choice *must* be made, and it must be made soon. All the subsidiary work, all the related enterprises, depend upon the fundamental difference in type. Opinions differ as widely to-day as they did at the time when the project was first considered by the international committee in 1879. Engineers of the highest standing at home and abroad have expressed themselves for or against one type or the other, but it is a question upon which no complete agreement is possible. In theory a sea-level canal has unquestionable advantages, but, practically, the elements of cost and time necessary for the construction preclude to-day, as they did in 1894, when the New Canal Company recommenced active operations, the building of a sea-level canal. It is *not a question of the ideally most desirable, but of the practically most expedient*, that confronts the American people and demands solution.

The New Panama Canal Company had approved the lock plan, which placed the minimum elevation of the summit level at 97.5 feet above the sea and the maximum level at 102.5 feet above the same datum. In the words of Prof. William H. Burr:

It provided for a depth of 29.5 feet of water and a bottom width of canal prism of about 98 feet, except at special places, where this width was increased. A dam was to be built near Bohio, which would thus form an artificial lake, with its surface varying from 52.5 to 65.6 feet above the sea. The location of this line was practically the same as that of the old company. The available length of each lock chamber was 738 feet, while the available width was 82 feet, the depth in the clear being 32 feet 10 inches. The lifts were to vary from 26 to 33 feet. It was estimated that the cost of finishing the canal on this plan would be $101,850,000, exclusive of administration and financing.

The Isthmian Commission of 1899–1901 considered the project, reëxamined into the facts, and as stated by Professor Burr—

The feasibility of a sea-level canal, but with a tidal lock at the Panama end, was carefully considered by the Commission, and an approximate estimate of the cost of completing the work on that plan was made. In round numbers this estimated cost was about $250,000,000, and *the time required to complete the work would probably be nearly or quite twice that needed for the construction of a canal with locks*. The Commission therefore adopted a project for the canal locks. Both plans and estimates were carefully developed in accordance therewith.

Professor Burr, now in favor of a sea-level canal, *then* concurred in the report in favor of a lock canal.

Since the Panama canal became the property of the nation a vast amount of necessary and preliminary work has been done preparatory to the actual construction of the canal. A complete civil government of the Canal Zone has been established, an army of experts and engineers has been organized, the work of sanitation and police control is in excellent hands, and the Isthmus, or, more properly speaking, the Canal Zone, is to-day in a better, cleaner, and more healthful condition than at any previous time in its history. A considerable amount of excavation and necessary improvements in transportation facilities have been carried to a point where further work must stop until the Isthmian Commission knows the final plan or type of the canal. The reports which have been made of the work of the Commission during its two years of actual control are a complete and affirmative answer to the question whether what has been done so far has been done wisely and well, and the facts and evidence prove that the present state of affairs on the Isthmus is in all respects to the credit of the nation.

Now, it is evident that the question of plan or type of canal is largely one for engineers to determine, but even a layman can form an intelligent opinion, without entering into all the details of so complex a problem as the relative advantage or disadvantage of a sea-level versus a lock canal. This much, however, is readily apparent, that a sea-level canal will cost a vast amount of money and may take twice the time to build, while it will not necessarily accommodate a larger traffic or ships of a larger size. A lock canal can be built which will meet all requirements; it can be built deep enough and wide enough to accommodate the largest vessels afloat; it can be so built that transit across the Isthmus can be effected in a reasonably short period of time—in a word, it is a practical project, which will solve every pending question involved in the construction of a transisthmian canal in a practical way, at a reasonable cost, and within a reasonable period of time.

To determine the question the President appointed an International Board of Consulting Engineers. The Board included in its membership the world's foremost men in engineering science, and the report is without question a most valuable document. The President, in his address to the members of the Board on September 11, 1905, outlined his views with regard to the desirability of a sea-level canal, if such a one could be constructed at a reasonable cost within a reasonable time. He said—

If to build a sea-level canal will but slightly increase the risk and will take but little longer than a multilock high-level canal, this, of course, is preferable. But if to adopt the plan of a sea-level canal means to incur great hazard and to incur indefinite delay, then it is not preferable.

The problem as viewed by the American people could not be more concisely stated. Other things equal, a sea-level canal, no doubt, would be preferable; but it remains to be shown that such a canal would in all essentials provide safe, cheap, and earlier navigation across the Isthmus than a lock canal.

For, as the President further said on the same occasion, there are two essential considerations: First, the greatest possible speed of construction; second, the practical certainty that the proposed plan will be feasible; that it can be carried out with the minimum risk; and in conclusion that—

There may be good reason why the delay incident to the adoption of a plan for an ideal canal should be incurred; but if there is not, then I hope to see the canal constructed on a system which will bring to the nearest possible date in the future the time when it is practicable to take the first ship across the Isthmus—that is, which will in the shortest time possible secure a Panama waterway between the oceans of such a character as to guarantee permanent and ample communication for the greatest ships of our Navy and for the largest steamers on either the Atlantic or the Pacific. The delay in transit of the vessels owing to additional locks would be of small consequence when compared with shortening the time for the construction of the canal or diminishing the risks in the construction. In short, I desire your best judgment on all the various questions to be considered in choosing among the various plans for a comparatively high-level multilock canal, for a lower-level canal with fewer locks, and for a sea-level canal. Finally, I urge upon you the necessity of as great expedition in coming to a decision as is compatible with thoroughness in considering the conditions.

The Board organized and met in the city of Washington on September 1, 1905, and on the 10th of January, 1906, or about four months later, made its final report to the President through the Secretary of War. The Board divided upon the question of type for the proposed canal, a majority of eight—five foreign engineers and three American engineers—being in favor of a canal at sea-level, while a minority of five—all American engineers—favored a lock canal at a summit level of eighty-five feet. The two propositions require separate consideration, each upon its own merits, before a final opinion can be arrived at as to the best type of a waterway adapted to our needs and requirements under existing conditions.

Upon a question so involved and complex, where the most eminent engineers divide and disagree, a layman can not be expected to view the problem otherwise than as a business proposition which, demanding solution, must be disposed of by a strictly impartial examination of the facts. Weighed and tested by practical experience in other fields of commercial enterprise, it is probably not going too far to say, as in fact it has been said, that there is entirely too much mere engineering opinion upon this subject and not a well-defined concentrated mass of data and solid convictions. It is equally true, and should be kept in mind, that the time given by the Board to the consideration of the subject in all its practical bearings, including an examination of actual conditions on the Isthmus, was limited to so short a period that it would be contrary to all human experience that this report should represent an infallible or final verdict for or against either of the two propositions.

It is necessary to keep in mind certain facts which may be concisely stated, and which I do not think have been previously brought to the attention of Congress. While the Board had been appointed by the President on June 24, 1905, the first business meeting did not take place until September 1st, and the final meeting of the full Board occurred on November 24th of the same year. This was the twenty-seventh meeting during a period of eighty-five days, after which there were three more meetings of the American members, the last having been held on January 31, 1906. Thus the actual proceedings of the full Board were condensed into twenty-seven meetings during less than three months, a part of which time—or, to be specific, six days—was spent on the Isthmus.

The minutes of the proceedings have been printed and form a part of the final report made to the President under date of January 10, 1906. They do not afford as complete an insight into the business transactions of the Board as would be desirable, and the evidence is wanting that the subject was as thoroughly discussed in all its details, with particular reference to the two propositions of a sea-level or a lock canal, as would seem necessary. Very important features necessary to the sea-level plan were treated in the most superficial way, guessed at, or wholly ignored. I do not hesitate to say that no banking house in the world called upon to provide funds necessary for an enterprise of this magnitude as a private undertaking would advance a single dollar upon a project as it is here presented by the majority of the Board to the American Congress as the final conclusion of engineers of the highest standing. The Board, as I have said, divided upon the question, and by a majority of eight pronounced in favor of a sea-level against a minority of five in favor of a lock canal. Let us inquire how this conclusion, of momentous importance to the nation, was arrived at and whether the minutes of the Board furnish a conclusive answer.

As early as the sixth meeting, or on September 16th—that is, after the Board had been only fifteen days in existence—a resolution was introduced by Mr. Hunter, chief engineer of the Manchester Ship Canal, requesting that a special committee be appointed to prepare at once a project for a sea-level canal.

Mr. Spooner. — What was the date of the resolution with respect to the lock canal?

Mr. Dryden. — October 3d, seventeen days afterwards.

In marked contrast, it was not until after the Board had visited the Isthmus and while the members were on their way home — that is, at sea — on October 3d, that, on motion of Mr. Stearns, a corresponding committee was appointed to prepare plans for a lock canal. The recital of dates is of very considerable importance, for it is evident that there was a decided and early preference on the part of certain members of the Board for a sea-level canal, and that to this particular project more attention was given and a more determined attempt was made to secure data in its defense than to the corresponding project for a lock canal. That is to say, while the special committee for the consideration of a sea-level canal had been appointed on September 16th, the corresponding committee to consider the lock project was not appointed until October 3d, or seventeen days later, with the additional disadvantage of the Board being on the ocean, with no opportunity to send for persons and papers during the short period of time remaining to take into due consideration all the facts pertaining to a lock canal, for, as I have said before, the last business meeting was held on November 24th.

Mr. Foraker. — Mr. President — —

The Vice President. — Does the Senator from New Jersey yield to the Senator from Ohio?

Mr. Dryden. — Certainly.

Mr. Foraker. — I would like to ask the Senator whether on the 16th of September, when this motion was made by Mr. Hunter, if I remember correctly, the Board of Engineers had completed their investigations and explorations on the Isthmus? I did not observe.

Mr. Dryden. — No.

Mr. Kittredge. — Mr. President — —

The Vice President. — Does the Senator from New Jersey yield to the Senator from South Dakota?

Mr. Dryden. — I yield.

Mr. Kittredge. — If the Senator from New Jersey will permit me, I will be glad to answer the question of the Senator from Ohio. The Board of Consulting Engineers sailed from New York on the 28th of September for the Isthmus and returned about the middle or 20th of October.

Mr. Foraker. — Sailed from the Isthmus?

Mr. Kittredge. — Sailed from New York for the Isthmus.

Mr. Foraker. — Then the motion was made by Mr. Hunter before the Board of Engineers left the United States.

Mr. Kittredge. — Certainly; to appoint a committee of investigation.

Mr. Dryden. — I should like to say at this point that while I have gladly yielded to Senators, I think it is quite probable that before I get through I shall cover any questions that may be asked. I would prefer to complete my remarks, and then I shall be very glad to answer any questions that Senators may choose to ask.

Mr. Foraker. — I beg pardon.

Mr. Dryden. — I was glad to yield to the Senator.

Mr. Foraker. — The speech is a very interesting one.

Mr. Dryden. — There is nothing in the minutes of the Board which disclosed that either proposition received the necessary deliberate consideration of the extremely complex and important details entering into the two respective projects, but it is evident that, regarding the sea-level proposition at least, there was a decided bias practically from the outset, which matured in the majority report favoring that proposition. What was in the minds of the members, what was done outside of the Board meetings, by what means or methods conclusions were reached, has not been made a matter of record and is not, therefore, within the knowledge of Congress.

It is true that the respective reports of the two committees were brought before the Board as a whole on November 14th and that the subject was discussed at some length on November 18th, when each member of the Board expressed his views for or against one of the two projects. But there remained only ten days before the last business meeting of the Board was held, when the foreign members sailed for home. The final reports, as they are now before Congress, apparently never received the proper and extended consideration of the Board as a whole, and the minority report favoring a lock canal seems never to have been discussed upon its merits at all. When I recall the very different procedure of the technical commission appointed by the New Panama Canal Company, which extended its consideration of the subject from February 3, 1896, to September 8, 1898, during which time ninety-seven stated meetings and a large number of informal meetings were held, I say, it seems to me, from a practical business point of view, casting no reflection upon either the ability or the fairness of judgment of the members of the International Board, that the mere element of time should weigh decidedly in favor of the verdict of the technical commission of 1898, which was unanimous for a lock canal.

Of the technical commission of 1896–1898, Mr. Hunter, chief engineer of the Manchester Ship Canal, was a member, and he at that time, without a word of dissent, joined the other members in giving the unanimous and emphatic expression of the committee in favor of a lock canal.

Mr. Teller.—Mr. President——

The Vice President.—Does the Senator from New Jersey yield to the Senator from Colorado?

Mr. Dryden.—Certainly.

Mr. Teller.—Will the Senator kindly repeat the date of that?

Mr. Dryden.—Of the technical commission of 1896–1898, Mr. Hunter, the chief engineer of the Manchester Canal, was a member. The technical commission was of the new French company.

Mr. Teller.—You refer to the commission of the new French company?

Mr. Dryden.—Yes, sir; the commission of the new French company.

Why he should now change his views and convictions and why he should now be so emphatic and pronounced in favor of a sea-level project is not set forth in anything that has been printed or been communicated to the Senate Committee on Interoceanic Canals. This hurried action, this scanty consideration, as I have stated, is the foundation upon which the advocates of the sea-level plan rest their appeal for support. This is the report and the evidence upon which Congress is requested to pronounce in favor of a sea-level project and give its indorsement to a plan which will involve the country in at least $100,000,000 of additional expenditure and which will delay the opening of the canal for practical purposes of navigation possibly for ten years or more after the lock canal can be finished and opened for use.

The Isthmian Commission restates certain points in a clear and precise way, which leaves no escape from the conclusion that both as to time and cost the majority members of the Board materially underestimated important factors, and that they have every reason to believe that the total estimate of cost of a sea-level canal should be raised to $272,000,000, and that the estimate of time for construction should be increased to at least fifteen and a half years. But under certain readily conceivable conditions it is practically certain that the construction of a sea-level canal will consume not less than twenty years.

The Isthmian Commission reëxamined carefully the question of relative efficiency of the proposed sea-level canal compared with a lock canal, and they pronounce emphatically and unequivocally in favor of the lock project. They consider that the assumed danger from accidents to locks by passing vessels or otherwise is greatly exaggerated, and hold that while no doubt accidents may occur, and possibly will occur, such dangers can and will be sufficiently guarded against by an effective method of supervision and control. They hold that a lock canal properly constructed and managed is in no sense a menace to the safety of vessels, and that much practical experience and particularly the half-century of successful operation of the "Soo" Canal have demonstrated the contrary beyond dispute. They point out that the canal with locks at a level of eighty-five feet will be a waterway three times the size, in navigable area, of the projected sea-level canal, and, omitting the locks from consideration, will therefore afford three times the shipping facilities.

They show that in the sea-level canal there will be many and serious curves, while in the lock canal the courses are straight and changes of direction will be made at intersecting tangents, the same as in our river navigation, in which serious accidents are practically unknown. They show that the courses in a lock canal can be marked with ranges which will greatly facilitate navigation, particularly at night. The Commission points out that the argument of the majority of the Board, that locks will limit the traffic capacity of the canal, carries very little if any weight, and they refer to the experience of the "Soo" Canal, through which there passes annually a larger traffic than through all the other ship canals of the world combined.

Finally, the Isthmian Commission discusses the cost of operation and maintenance. The majority of the Board submit no details upon this most important item in canal construction and subsequent operation. What banking house in the world would advance a single dollar upon a canal or railway project upon a mere statement of the probable ultimate cost, but with no corresponding information as to cost of maintenance and operation! Having been appointed to reëxamine into all the facts, and, so to speak, to reconsider the entire project, the majority seriously erred in omitting from their report the necessary data and calculations for an accurate and trustworthy estimate of the cost of operation and maintenance of a sea-level canal.

From this point of view and in the light of the facts as presented by the Board for or against either project, the Isthmian Commission could not consistently act otherwise than to give their final approval to the more specific and practical recommendations of the minority members of the Board, and they properly say that *"it appears that the canal proposed by the minority of the Board of Consulting Engineers can be built in half the time and for a little more than half of the cost of the canal proposed by the majority of the Board."* They advance a number of specific reasons why a lock canal when completed will for all practical purposes—commercial, military, and naval—be a better canal than a sea-level waterway with a tidal lock, as proposed by the majority members of the Board.

The report of the Board was carefully and critically examined by Chief Engineer Stevens, of the Isthmian Commission and in actual charge of engineering matters on the Isthmus. Mr. Stevens is a man of very large practical American engineering experience, and he adds to the finding of the Commission the weight of his authority, decidedly and unequivocally in favor of a lock canal. He states as the sum of his conclusions that, all things considered, the lock or high-level canal is preferable to the sea-level type, so-called, for the reason that it will provide a safer and quicker passage for ships; that it will provide beyond question the best solution of the vital problem of how safely to care for the flood waters of the Chagres and other streams; that provision is offered in the lock project for enlarging its capacity to almost any extent at very much less expense of time and money than can be provided for by any sea-level plan; that its cost of operation, maintenance, and fixed charges, including interest, will be very much less than any sea-level canal, and that the time and cost of its construction will not be more than one-half that of a canal of the sea-level type; that the lock project will permit of navigation by night; and that, finally, even at the same cost in time and money, Mr. Stevens would favor the adoption of the high-level lock canal plan in preference to that of the proposed sea-level canal.

To these observations and comments the Secretary of War, under whose supervision this great work is going on, adds his opinion, which is decidedly and unequivocally in favor of a lock canal. In his letter to the President, Mr. Taft goes into all the important details of the subject and reveals a masterly grasp of the situation as it confronts the American people at the present time. He calls attention to the fact that lock navigation is not an experiment; that all the locks in the proposed canal are duplicated, thereby minimizing such dangers as are inherent in any canal project, and he adds that experience shows that with proper plans and regulations the dangers are much more imaginary than real. He goes into the facts of the proposed great dam to be constructed at Gatun and points out that such construction is not experimental, but sustained by large American experience, which is larger, perhaps, than that of any other country in the world. He gives his indorsement to the views of the Isthmian Commission and its chief engineer that the estimated cost of time and money for completing a sea-level canal is not correctly stated by the majority members of the Board, and that the cost, in all probability, will be at least $25,000,000 more, while, in his opinion, eighteen to twenty years will be necessary to complete the sea-level project. He also holds that the military advantages will be decidedly in favor of a lock canal.

This is practically the present status of facts and opinions regarding the canal problem as it is now before Congress, except that since January the Senate Committee on Interoceanic Canals has collected a large mass of additional and valuable testimony. Restating the facts in a somewhat different way, Congress is asked to give its final approval to the sea-level proposition, chiefly favored by foreigners, and to give its disapproval to the project of a lock canal, favored by American engineers. Congress is asked to rely in the main upon the experience gained in the management of the Suez Canal, where the conditions are essentially and fundamentally different from what they are or ever will be on the Isthmus of Panama, and to disregard the more than fifty years' experience in the successful management of the lock canals connecting the Great Lakes. Congress is asked to pronounce against the lock canal because in the management of the ship canal at Manchester several accidents have occurred, due to carelessness or ignorance in navigation, and we are asked to disregard the successful record of the "Soo" Canal, in the management of which only three accidents, of no very serious importance, have occurred during more than fifty years.

In no other country in the world has there been more experience with lock canals than in this. For nearly a hundred years the Erie Canal has been one of our most successful of inland waterways, connecting the ocean with the Great Lakes. The Erie Canal is 387 miles in length, has 72 locks, and is now being enlarged, to accommodate barges of a thousand tons, at a cost of $101,000,000. We have the Ohio Canal, with 150 locks; the Miami and Erie Canal, with 93 locks; the Pennsylvania Canal, with 71 locks; the Chesapeake and Ohio Canal, with 73 locks; and numerous other inland waterways of lesser importance. It is a question of degree and not of kind, for the problem is the same in all essentials, and confronts Congress as much in the proposed deep waterway connecting tide-water with the Great Lakes, in which locks are proposed with a lift of 40 feet or more, or very considerably in excess of the proposed lift of the locks on the Isthmian Canal.

The proposed ship canal from Lake Erie to the Ohio River provides for 34 locks. The suggested canal from Lake Michigan to the Illinois and Mississippi rivers provides for 37 locks, and, finally, the projected ship canal from the St. Lawrence River to Lake Huron contemplates 22 locks. So that lock canals of exceptional magnitude are not only in existence, but new canals of this type are contemplated in the United States and Canada.

In other words, Congress is asked to regard with preference the judgment and opinions of foreign engineers and to disregard the judgment and opinions of American engineers. We are seriously asked to completely disregard American opinion, as voiced by the Isthmian Commission, responsible for the enterprise as a whole; as voiced by the Secretary of War, responsible for the time being for the proper execution of the work; as voiced by Chief Engineer Stevens, who stands foremost among Americans in his profession; and finally, as voiced by all the engineers now on the Isthmus, who have a practical knowledge of the actual conditions, and who are as thoroughly familiar as any class of men with the problems which confront us and with the conditions which will have to be met. I for one, leaving out of consideration for the present details which are subject to modification and change, believe that it will be a fatal error for the nation to commit itself to the practically hopeless and visionary sea-level project and to delay for many years the opening of this much needed waterway connecting the Atlantic with the Pacific. I for one am opposed to a waste of untold millions and to additional burdens of needless taxation, while the project of a lock canal offers every practical advantage, offers a canal within a reasonable period of time and at a reasonable cost, offers a waterway of enormous advantage to American shipping, of the greatest possible value to the nation in the event of war, and the opportunity for the American people to carry into execution at the earliest possible moment what has been called the "dream of navigators," and what has thus far defied the engineering skill of European nations.

But in addition to the evidence presented for or against a sea-level or lock canal project by the two conflicting reports of the Board of

Consulting Engineers, there is now available a very considerable mass of testimony of American engineers who were called as witnesses before the Senate Committee on Interoceanic Canals. The testimony has been printed as a separate document and makes a volume of nearly a thousand pages. Much of this evidence is conflicting, much of it is mere engineering opinion, much of it comes perilously near to being engineering guesswork, but a large part of it is of practical value and may safely be relied upon to guide the Congress in an effort to arrive at a final and correct conclusion respecting the type of canal best adapted to our needs and requirements.

A critical examination and review of this testimony, as presented to the Senate Committee from day to day for nearly five months, including the testimony of administrative officers and others, relating to Panama Canal affairs generally, is not practicable at this stage of the session. Among others, the committee examined Mr. John F. Stevens, chief engineer, upon all the essential points in controversy, regarding which, in the light of additional experience and a very considerable amount of new and more exact information, Mr. Stevens reaffirms his convictions in favor of the practicability and superior advantages of a lock canal.

In opposition to the views and conclusions of Mr. Stevens, Prof. William H. Burr pronounced himself emphatically in favor of the sea-level project. As a member of the former Isthmian Commission, reporting upon the type of canal, Mr. Burr had signed the report in favor of the lock project, but as a member of the Board of Consulting Engineers he had sided with the majority favoring the sea-level canal. Thus engineering opinion is as apt as any other human opinion to undergo a change, and the convictions of one year in favor of a proposition may change into opposite convictions, favoring an opposite proposition, only a few years later. Mr. William Barclay Parsons, also a member of the Board of Consulting Engineers, who had signed the report in favor of the sea-level project, gave further evidence before the committee, restating his views and convictions in favor of the sea-level type. Mr. William Noble, an engineer of large experience, for some years in charge of the "Soo" Canal, and who, as a member of the Board of Consulting Engineers, had signed the report in favor of a lock project, restates his views and convictions in favor of a lock canal. Mr. Noble had also been a member of the Isthmian Commission of 1902, reporting at that time in favor of a lock canal.

Mr. Frederick P. Stearns, the foremost American authority on earth-dam construction, gave evidence regarding the safety of the proposed dams at Gatun and other points. His views and conclusions are based upon large practical experience and a profound theoretical knowledge of the subject. Mr. Stearns had also been a member of the Consulting Board of Engineers and as such had signed the report of the minority in favor of the lock project. He reaffirmed his views favoring a lock canal with a dam at Gatun. Mr. John F. Wallace, former chief engineer, gave testimony in favor of the sea-level type and strongly opposed the lock project. Col. Oswald H. Ernst, United States Army, than whom probably few are more thoroughly familiar with conditions on the Isthmus and the entire project of canal construction, declared himself to be strongly in favor of the lock-canal project.

Gen. Peter C. Hains, United States Army, equally well qualified to express an opinion on the subject in all its important points, pronounced himself strongly and unequivocally in favor of a lock canal.

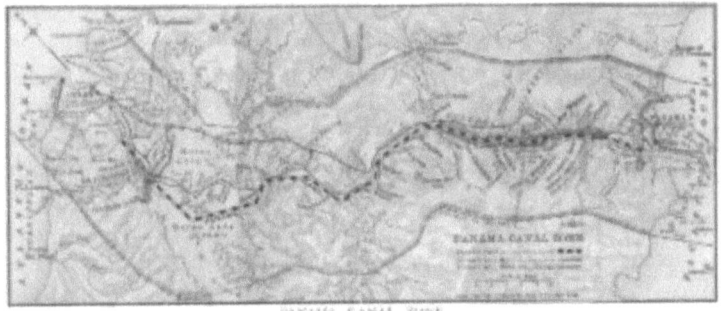

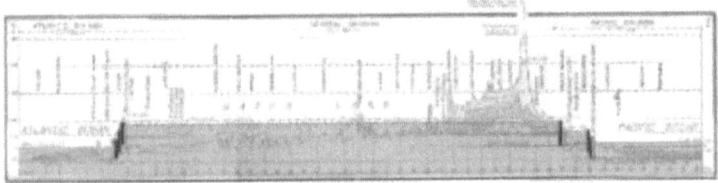

PANAMA CANAL ZONE

PROFILE OF CANAL

Gen. Henry L. Abbot, United States Army, one of the highest authorities on river hydraulics, thoroughly familiar with Mississippi River flood problems, a former member of the International Technical Commission, of the New Panama Canal Company, and for a time its consulting engineer, a member of different Isthmian commissions, and also a member of the consulting board, reëmphasized his conviction, sustained by much valuable evidence, in favor of the lock-canal project. General Abbot, as a member of the consulting board, had signed the report of the minority in favor of a lock canal. Gen. George W. Davis, United States Army, for a time governor of the Canal Zone and president of the International Board of Consulting Engineers, restated his views and convictions as opposed to the lock-canal type and in favor of the sea-level project. The last witness, Mr. B.M. Harrod, an engineer of large experience, for many years connected with levee construction and familiar with the flood problems of the Mississippi River, submitted a statement in which he restated his views in favor of a lock canal.

So that, summing up the evidence of twelve engineers examined before the committee (including Mr. Lindon W. Bates), there were eight American engineers strongly and unequivocally in favor of a lock canal, while four expressed their views to the contrary. Subjecting the mass of testimony to a critical examination, I cannot draw any other conclusion or arrive at any other conviction than *that the lock project, in the light of the facts and large experience, has decidedly the advantage over the sea-level proposition.* And this view is strengthened by the fact that the opinion of the engineers most competent to judge—that is, men like Mr. Noble, who has thoroughly studied lock-canal construction, management, and navigation, who as a member of the United States Deep Waterway Commission reëxamined probably as thoroughly as any living authority into the entire subject of the mechanics and practice of lock canals—is emphatically opposed to the sea-level proposition.

When a man like Mr. Stearns, of national and international reputation as a waterworks engineer, who for many years has been in charge of the extensive construction work of the Massachusetts Metropolitan Water and Sewerage Board, and who probably has as large a practical and theoretical knowledge of earth-dam construction as any living authority, declares himself to be strongly in favor of the lock project and believes in the entire safety of the dams required in connection therewith, I hold that such a judgment may be relied upon and that it should govern in national affairs as it would govern in private affairs if the canal construction were a business enterprise and involved the risk of private capital. When we find a man like Mr. Harrod, who for many years has been in charge of levee construction in Louisiana, thoroughly familiar with the theory and practice of river and flood control, express himself in favor of the lock project and in opposition to the sea-level canal, I hold that we may with entire confidence accept his judgment as a governing principle in arriving at a final decision respecting the type of the canal to be finally fixed by the Congress.

And, going back to the minority report of the Board of Consulting Engineers, we find that Mr. Joseph Ripley, the general superintendent at present in charge of the "Soo" Canal, and Mr. Isham Randolph, chief engineer of the sanitary district of Chicago, and thoroughly familiar with canal construction and management, both

American engineers of much experience and high standing, pronounce themselves in favor of a lock canal. When confronted by these facts, I for one would rely upon American engineers, American conviction and American experience, and accept the lock-canal proposition.

In this matter, as in all other practical problems, we may safely take the business point of view, and calculate without bias or prejudice the respective advantages and disadvantages; and the more thorough the method of reasoning and logic applied to the canal problem the more emphatic and incontrovertible the conclusion that the Congress should decide in favor of a plan which will give us a navigable waterway across the Isthmus within a measurable distance of time and with a reasonable expenditure of money, as opposed to a visionary theory of an ideal canal which may ultimately be constructed, possibly for the exclusive benefit of future generations, but at an enormous waste of money, time, and opportunity. I do not think we want to repeat at this late stage of the canal problem the fatal error of De Lesseps, who, when he had the opportunity in 1879 to make a choice of a practical waterway, being influenced by his great success at Suez, upon the most fragmentary evidence and in the absence of definite knowledge of actual conditions, decided beforehand in favor of a sea-level canal. It was largely his bias and prejudice which proved fatal to the enterprise and to himself.

I may recall that the so-called "international congress of 1879" was a mere subterfuge; that the opinions of eminent engineers, including all the Americans, were opposed to a sea-level project and in favor of a lock canal, but De Lesseps had made his plans, he had arrived at his decision, and in his own words, at a meeting of the American Society of Civil Engineers held in January, 1880, said, "I would have put my hat on and walked out if any other plan than a sea-level canal project had been adopted."

The situation to-day is very similar to the critical state of the canal question in 1902. What was then a question of choice of route is to-day a question of choice of plan. What was then a geographical conflict is to-day a conflict of engineering opinions. It has been made clear by the reference to the report of the Board of Consulting Engineers and by the testimony of the engineers before the Senate committee that the opinion of eminent experts is so widely at variance that there is little, if any, hope of an ultimate reconciliation. It is a choice of one plan or the other—of a sea-level or a lock canal. In respect to either plan a mass of testimony and data exists, which has been brought forward to sustain one view or the other. In respect to either plan there are advantages and disadvantages. The majority of the Senate Committee on Interoceanic Canals have reported favorably a bill providing for the construction of a canal at sea level. From this majority opinion the minority of the committee emphatically and unequivocally dissent, and in their report they express themselves in favor of the lock canal.

The minority report calls attention to the changed conditions and requirements, which now demand a canal of much larger dimensions than originally proposed. Even as late as 1901 the depth of the canal prism was to be only 35 feet, against 40 to 45 feet in the project of only five years later. The bottom width has been increased from 150 to 200 feet and over. The length of the locks has been changed from 740 to 900 feet, and the width from 84 to 90 feet. These facts must be kept in mind, for they bear upon the questions of time and cost, and a sea-level or lock canal, as proposed to-day, is in all respects a very much larger affair, demanding very superior facilities for traffic, than any previous canal project ever suggested or proposed. This change in plans was made necessary by the Spooner act, which provides for a canal of such dimensions that the largest ship now building, or likely to be built within a reasonable period of time, can be accommodated.

Now, the estimated saving in money alone by adopting the lock plan—that is, on the original investment, to say nothing of accumulating interest charges—would be at least $100,000,000. Granting all that is said in favor of a sea-level canal, it is not apparent by any evidence produced that such a canal would prove a material advantage over a lock canal. All its assumed advantages are entirely offset by the vastly greater cost and longer period of time necessary for construction, and I am confident that they would not be considered for a moment if the canal were built as a commercial enterprise. I do not think that they should hold good where the canal is the work of the nation, because a vast sum of money otherwise needed will be eventually sunk if the sea-level project is adopted, and entirely upon the theory that if certain conditions should arise *then* it would be better to have a sea-level than a lock canal. We have never before proceeded in national undertakings upon such an assumption; we have never before, as far as I know, deliberately disregarded every principle of economy in money and time; we have never before in national projects attempted to conform to merely theoretical ideas, but we have always adhered to practical, hard common-sense notions of *what is best* under the circumstances.

The majority of the committee attack the proposition that the proposed lock canal will have "locks with dimensions far exceeding any that have ever been made." If this principle were adopted in every other line of human effort all advancement would come to an end—even the canal enterprise itself—for, as it stands to-day, it far exceeds in magnitude any corresponding effort ever made by this or any other nation. They say that the proposed flight of three locks at Gatun would be objectionable and unsafe, but we have the evidence of American engineers of the highest standing, whose reputations are at stake, who are absolutely confident that these locks can be constructed and operated with entire safety. The committee say that "the entry through and exit from these contiguous locks is attended with very great danger to the lock gates and to the ships as well"; but if mere inherent danger of possible accidents were an objection there would be no great steamships, no great battleships, no great bridges and tunnels, no great undertakings of any kind.

The committee point out that accidents have occurred in the "Soo" Canal and in the Manchester Ship Canal; but the conditions, in the first place, were decidedly different, and, in the second place, they proved of no serious consequence as a hindrance to traffic and did no material injury to the canal. The "Soo" Canal has been in operation as a lock canal for some fifty years; it has been enlarged from time to time, and to-day accommodates a larger traffic than passes through all other ship canals of the world combined. It is a sufficient answer to the objections to say that this experience should have a determining influence in arriving at a conclusion, for the inherent problems of lock-canal construction are as well understood by American engineers as any other problems or questions in engineering science. The proposed deep waterway with a 30-foot channel from Chicago to tide-water, which has been surveyed by direction of Congress, proposes an expenditure of $303,000,000, and several locks with a lift of 40 feet or more. The enlargement of the Erie Canal by the State of New York, at an expenditure of $101,000,000, involves engineering problems, including lock construction, not essentially different from those inherent in the lock-canal project at Panama; and if these problems can be solved by our engineers at home, it stands to reason that we may rely upon their judgment that they can be solved at Panama.

The majority of the Senate committee object to the proposed dam at Gatun, and say that—

Earth dams founded on the drift and silt of ages, through which water habitually percolates, to be increased by the pressure of the 85-foot lock when made, have been referred to by many of our technical advisers as another element of danger. The vast masses of earth piled on this alluvial base to the height of 135 feet will certainly settle, and as the drift material of this base or foundation has varying depth, to 250 feet or more, the settlement of the new mass, as well as its base, will be unequal, and it is predicted that cracks and fissures in the dam will be formed, which will be reached and used by the water under the pressure above mentioned, and will cause the destruction of the dam and the draining off of the great lake upon which the integrity of the entire canal rests.

But all of this is mere conjecture. The evidence of Engineer Stearns, a man of large experience, and of Engineer Harrod, familiar with river hydraulics and levee construction, and of many others, is emphatically to the contrary. There is not an American engineer of ability, nor an American contractor of experience, who would not undertake to build the proposed dam at Gatun and guarantee its safety and permanency without any hesitation whatever. The alternative proposal of a dam at Gamboa would be as objectionable upon much the same ground, and the dam there, which is indispensable to the sea-level project, has also been considered unsafe by some of the engineers. In all questions of this kind the aggregate experience of mankind ought to have greater weight than the abstract theories of individuals, and I am confident that our engineers, who have so successfully solved problems of the greatest magnitude in the reclamation projects of the far West and in the control and regulation of the floods of the Mississippi River, will solve with equal success similar problems at Panama.

The committee further says that the sea-level project contemplates the removal of some 110,000,000 cubic yards of material, while the lock canal would require the removal of only about half that quantity, or, in other words, that there is a difference of some 57,000,000 cubic yards, which, "to omit to take out ... is to confess our impotence, which is not characteristic of the American people or their engineers or contractors." By this method of reasoning a nation which can build a battleship of 16,000 tons displacement is impotent if it can not build one of twice that tonnage, and if this reason applies to quantity of material, why not say that a nation which can dig a canal 150 feet wide through a mountain some seven miles in length admits its impotence if it can not dig one 300 feet wide, or 600 feet, if it should please to do so? But why should it be less difficult or a declaration of impotency on the part of our engineers to build a safe lock canal including a satisfactory and safe controlling dam at Gatun? As I conceive the problem, it is one of reasonable compromise, and while I do not question the ability of American engineers and contractors to build a sea-level canal, I am convinced by the facts in evidence that they can not do it within the time and for the money assumed by the advocates of the sea-level project.

This question of *time* is of supreme importance. Ten years in a nation's life is often a long space in national history. Many times the map of the world has been changed in less than a decade. No man in 1890 anticipated the war with Spain in 1898, and no man in 1906 can say what important event may not happen before the next decade has passed. The progress during peace is far greater in its permanent effect than the changes brought about by war. The world's commerce, the social, commercial, and political development of the South American republics and of Asiatic nations, all depend, more or less, upon the completion of an Isthmian waterway. It is the duty of this nation, since we have assumed this task, to construct a waterway across the Isthmus within the shortest reasonable period of time. Valuable years have passed, valuable opportunities have gone by. In 1884 De Lesseps, with supreme confidence and upon the judgment of his engineers, anticipated the opening of the Panama Canal in 1888. That was nearly twenty years ago. Shall it be twenty years more before that greatest event in the world's commercial history takes place? Had De Lesseps in 1879 gone before the International Congress with a proposition for a feasible canal at reasonable cost, free from prejudice or bias, had he then adopted the American suggestion for a lock canal, he would probably have lived to see its completion, and the world for fifteen years would have had the use of a practical waterway across the Isthmus.

As to safety in operation, which the committee discuss in their report, there is one very important point to be kept in mind, and that is that nine-tenths, or possibly a larger proportion, of shipping will be of vessels of relatively small size. If this should be the case, then the sea-level project contemplates a canal chiefly designed to meet the possible needs of a very small number of vessels of largest size, while the lock canal provides primarily for the accommodation of the class of steamships which of necessity would make the largest practical use of the Isthmian waterway. Now, it stands to reason that special precautions would be employed during the passage of a very large vessel, either merchantman or man-of-war, and even if necessity should demand the rapid passage of a fleet of vessels, say twenty or thirty, it is not conceivable that a condition would arise which could not be efficiently safeguarded against by those in actual charge and responsible for safety in the management of the canal. Considering the immense tonnage passing through the "Soo" Canal, which would not be equaled in the Panama Canal for a century to come, the very few and relatively unimportant accidents which have occurred during the fifty years of operation of that waterway are in every respect the most suggestive indorsement of the lock-canal project which could be advanced.

The time of transit, in the opinion of the majority committee of the Senate, would be somewhat longer in the case of a lock canal. This may be so, though much depends upon the class of ships passing through and their number. To the practical navigator the loss of a few hours would be a negligible quantity compared with the higher tolls that will necessarily be charged if an additional $100,000,000 is expended in construction and an additional interest burden of at least $2,000,000 per annum has to be provided for. I understand that the actual value of an hour or two in the case of commercial ships of average size would be a matter of comparatively no importance in contrast with the all-suggestive fact that the alternative project of a sea-level canal would provide no navigation whatever across the Isthmus for probably ten years more. If it is an advantage to gain an hour or two in transit ten years hence by having no transisthmian shipping facilities for the ten years in the meantime, then it might as well be argued that it would be better to project a sea-level canal 300 feet wide at every point, so that the commerce of the year 2000 may be properly provided for. But to the practical navigator of the year 1916, who leaves the port of New York for San Francisco by way of Cape Horn, a possible loss of two or three hours or more would be many times preferable, if the Isthmus were open for traffic, to a certain loss of from forty to fifty days to make the voyage all around South America.

Upon the question of cost of maintenance the majority committee in their report point out that the Board of Consulting Engineers did not submit the details of any estimate of cost of maintenance, repairs, etc., but they say that this factor was properly taken into account by the minority favoring a lock canal. Now, there is probably no more important question connected with the whole canal problem than this, for if the annual expense of maintenance, to be provided for by Congressional appropriations, should attain such an exorbitant figure as to make any fair return upon the investment impossible, it is conceivable that the most serious political and financial consequences might arise and the success of the enterprise itself might be placed in jeopardy. Upon a maximum cost, in round figures, of $200,000,000 for a lock canal, and of $300,000,000 as a minimum for a sea-level canal, the additional annual interest charge would be at least $2,000,000.

But Mr. Stearns estimates that under certain conditions a sea-level canal might cost as much as $410,000,000, which would add millions of dollars more per annum to the fixed charges which must be included in the cost of maintenance, to say nothing of a possibly much higher cost of operation. Nor can I agree to the statement that the cost of operation of a sea-level canal would be $800,000 per annum less than in the case of a lock canal; but, on the contrary, I am fully satisfied that the expense would be very much greater in the sea-level project, if proper allowance is made for interest charges upon the additional outlay, which cannot be rightfully ignored.

Upon this important point the evidence of the engineers and of the minority members of the Board is strongly in favor of the lock-canal project.

As regards ultimate cost, the estimates of the majority are very much more indefinite and conjectural than the more carefully prepared estimates of the minority of the Board of Consulting Engineers. Upon this point the majority of the Senate committee say:

There are two estimates now before the Senate, both originating with the Board of Consulting Engineers. The basis of computation of cost at certain unit prices was adopted unanimously by the Board, and we are told that the cost, with the 20 per cent. allowance for contingencies, will be, for the sea-level canal, the sum of $247,021,200. Your committee has adopted the figures stated by the majority on page 64 of its report of a total of $250,000,000 for the ultimate final cost of the sea-level canal.

The estimate of the minority for a lock canal at a level of eighty-five feet is, in round figures, $140,000,000, or about $110,000,000 less than for a sea-level canal, which would represent a difference of $2,200,000 per annum in interest charges at the lowest possible rate of two per cent. The majority of the Senate committee attempt to meet this difference by capitalizing the estimated higher maintenance charge, which they fix at $800,000 per annum, and they thus increase the total cost of a lock canal by $40,000,000; but this, I hold, involves a serious financial error, unless a corresponding allowance is made for the ultimate cost of the sea-level project. There is, however, no serious disagreement upon the point that a sea-level canal in any event would cost a very much larger sum as an original outlay, certainly not less than $120,000,000 more, and, in all probability, in the opinion of qualified engineers, including Mr. Stevens, the chief engineer, twice that sum.

Reference is made in the report to the probable value of the land which will be inundated under the lock-canal project with a dam at Gatun, the value of which has been placed at approximately $300,000. The majority of the Senate committee estimate that this amount might reach $10,000,000, or as much as was paid for the entire Canal Zone. The estimate is based upon the price of certain lands required by the government near the city of Panama, but one might as well estimate the worth of land in the Adirondacks by the prices paid for real estate in lower New York. The item, no doubt, requires to be properly taken into account, but two independent estimates fix the probable sum at $300,000 for lands which are otherwise practically valueless and which would only acquire value the moment the United States should need them. In my opinion, the value of these lands will not form a serious item in the total cost of the canal, and I have every reason to believe that independent estimates of the minority engineers of the Consulting Board, and of Mr. Stevens, may be relied upon as conservative.

The majority of the Senate committee further say that—

It is not necessary to dwell upon the fact that all naval commanders and commercial masters of the great national and private vessels of the world are almost to a man opposed unalterably to the introduction of any lock to lift vessels over the low summit that nature has left for us to remove.

I am not aware that any material evidence of this character has come before the Senate Committee on Isthmian Affairs, investigating conditions at Panama. I do know this, however, that until very recently it has been the American project to construct a lock canal. All the former advocates of an American canal by way of Panama or Nicaragua, or by any other route, contemplated a lock canal of a much more complex character than the present Panama project. All the advocates of a canal across the Isthmus, including many distinguished engineers in the army and navy, have been in favor of a lock canal, and almost without exception have reported upon the feasibility of a lock canal across the Isthmus and upon its advantages to commerce and navigation, and in military and naval operations in case of war. The Nicaragua Canal, as recommended to Congress and as favored by the first Walker Commission, provided for a lock project far more complex than the proposition now under consideration.

Colonel Totten, who built the Panama railroad, recommended as early as 1857 the construction of a lock canal; Naval Commissioner Lull, who made a careful survey of the Isthmus in 1874, recommended a lock canal with a summit level of 124 feet and with 24 locks. Admiral Ammen, who, by authority of the Secretary of War, attended the Isthmian Congress of 1879, favored a lock project, in strong opposition to the visionary plan of De Lesseps. Admiral Selfridge and many other naval officers who have been connected with Isthmian surveying and exploration have never, to my knowledge, by as much as a word expressed their apprehensions regarding the feasibility or practicability of a lock canal.

As a matter of fact and canal history, the lock project has very properly been considered "an American conception of the proper treatment of the Panama canal problem." Mr. C.D. Ward, an American engineer of great ability, as early as 1879 suggested a plan almost identical with the one now recommended by the minority of the Consulting Board, including a dam at Gatun, instead of Bohio or Gamboa; and, in the words of a former president of the American Society of Civil Engineers, Mr. Welsh, "The first thought of an American engineer on looking at M. De Lesseps' raised map is to convert the valley of the lower Chagres into an artificial lake some twenty miles long by a dam across the valley at or near a point where the proposed canal strikes it, a few miles from Colon, such as was advocated by C.D. Ward in 1879." The site referred to was Gatun, and this was written in 1880, when the sea-level project had full sway.

So that it is going entirely too far to say that all naval commanders and commercial masters are in favor of the sea-level project. Admiral Walker himself, as president of the former Isthmian Commission, and as president of the Nicaraguan Board, favored a lock canal. Eminent army engineers, like Abbot, Hains, Ernst, and others, favor the lock project. It requires no very extensive knowledge of navigation to make it clear that passing through a waterway which for 35 miles, or 71 per cent. of its distance, will have a width of 500 feet or more, compared with one which, for the larger part, or for some forty-one miles, will have a width of only 200 feet or less, must appeal to the sense of security of the skipper while taking his vessel through the canal.

But it is a question of general principles, and not of personal preference. Our concern is with a matter of fact, and not with a theory. No ship-owner on the Great Lakes considers it a serious hindrance to navigation for vessels to pass through the lock of the "Soo" Canal;

no shipper running 1,000-ton barges through the future Erie Canal will have the least apprehension of danger or destruction; no captain navigating a vessel or boat through the proposed deep waterway from the ocean to the Lakes will hesitate to pass through locks with a proposed lift of over forty feet. These apprehensions are imaginary and not real. They are not derived from experience or from a summary statement of shipmasters and naval officers, but from the individual expressions and prejudice of a few who are opposed to the lock project. I am confident that if the matter is left to the practical navigator, to the ship-owner, and to the self-reliant naval officer, there will be no serious disagreement with the opinion that a lock canal, which can be built within a reasonable period of time, is preferable to any sea-level canal which may be built and opened to navigation twenty years hence or later.

There are two objections made by the majority of the Senate committee against a lock canal which require more extended consideration. These are, the protection of the canal in case of war and the danger of serious injury or total destruction by possible earth movements or so-called "earthquakes." Regarding the military aspects of the canal problem, the majority of the Senate committee say:

The Spooner act and the Hay-Varilla treaty contemplated the fortification and military protection of the canal route. No proposition affecting this policy is now before the Senate. In so far as the type of canal to be adopted has a bearing upon the jeopardy to or immunity of the canal from risk of malicious injury, the subject of safety and protection is pertinent and most important. If a canal of one type would be more liable to injury than another, this liability should under no circumstances be neglected in determining the type or plan. It does not require argument that the use of the canal by the United States will cease if the control passes to a hostile power between which and the United States a state of war exists, but this is true whatever the type may be.

As the majority of the committee point out, "no proposition affecting this project is now before the Senate." In my opinion, none is necessary. The neutrality of the canal is, by implication at least, assured, and we have pledged our national good faith that the waterway will be open to all the nations of the world. Some time in the future, when the canal is completed and an accepted fact, it may be advisable to adopt the course pursued in the case of the Suez Canal. The original concession for that canal provided, by section 3, for its subsequent fortification, but this was never carried into effect. By a convention dated December 22, 1888, among Great Britain, Germany, and other nations, the free navigation of the Suez Canal was made a matter of international agreement, and the same has been reprinted as Senate Document No. 151, Fifty-sixth Congress, first session, under date of February 6, 1900.

This, in any event, is a problem of the future. The canal is the property of the United States, and we shall always retain control. In the event of war we shall rely with confidence upon our navy to protect our interests on the Pacific and in the Caribbean Sea, but even more may we rely upon the all-important fact that it could never be to the interest of any other nation sufficient in size to be at war with us to destroy this international waterway, which will become an important necessity to the commerce of each and all. No neutral nation engaged in extensive commerce or trade would for an instant allow another nation at war with the United States to injure or destroy the canal or to seriously interfere with the traffic passing through it. To destroy as much as a single lock, to injure as much as a single gate, would be considered equal to an act of war by every commercial nation of the earth. In this simple fact lies a greater assurance of safety than in all the treaties which might be made or in all the fortifications which might be established to protect the canal.

The majority of the committee well say in their report, that the power of mischief "is within easy reach of all." The possibility of an assumed occurrence is very remote from its reasonable probability. We have to rely upon our own good faith and the watchful eyes of our officers. Against possible contingencies, such as are implied in the assumed destruction of the locks by dynamite or other high explosives, we can do no more than take the same precautions which we take in all other matters of national importance. We have to take our chances the same as any other nation would; the same as commercial enterprise would. Certainly the remote possibility of such an event, the still more remote contingency that the injury would be serious or fatal to the operation of the canal, should not govern in a decision to construct a canal for the use of the present generation rather than for the generations to come. No canal can be built free from vulnerable points; no forts, no battleships, can be built free from such a risk. It would be folly to delay the construction of a canal; it would be folly to sink a hundred million dollars or more upon so remote a contingency as this, which belongs to the realm of fanciful or morbid imagination rather than to the domain of substantial fact and actual experience.

As a last resort, the opposition to a lock canal brings forward the earthquake argument. It is a curious reminder of the early and bitter opposition to the building of the Suez Canal; its enemies had to fall back upon the absurd theory that the canal would prove a failure because the blowing sands of the desert would soon fill the channel. It was seriously proposed to erect a stone wall four feet high on each side of the embankment to provide against this imaginary danger to the canal. Another early objection to the Suez Canal was that the Red Sea level was 30 feet above the level of the Mediterranean, only set at rest in 1847 by a special commission, which included Mr. Robert Stephenson, the great son of a great father, bitter to the last in his opposition to the canal, which he considered an impracticable engineering scheme. There was much talk about the assumed prevalence of strong westerly winds on the southern Mediterranean coast, and the danger of constantly increasing deposits of the Nile, it was said, would render the establishment of a port impossible. It was necessary to place a war-ship for a whole winter at anchor three miles from the shore to prove the error of this assumption and set at rest a foolish rumor which came near proving fatal to the enterprise.

Earthquakes have occurred on the Isthmus, and there is record of one shock of some consequence in 1882. The matter has been inquired into in a general way by the various Isthmian commissions, and assumed some prominence during the discussions and debates regarding a choice of routes. It was plain to even the least informed that the volcanic belt of Nicaragua constituted a real menace to a canal in that region; and one of the strongest arguments advanced in the minority report of the Senate committee of 1902, submitted by Senator Kittredge, now a leading advocate of the sea-level project, in opposition to the Nicaragua Canal, was the assertion of the practical freedom of the Panama Isthmus from the danger of earth movements.

The minority of the Senate committee of 1902 in their report, summing up the final reasons in favor of the Panama route (section 12), said:

At Panama earthquakes are few and unimportant, while the Nicaraguan route passes over a well-known coastal weakness. Only five disturbances of any sort were recorded at Panama, all very slight, while similar official records at San Jose de Costa Rica, near the route of the Nicaragua Canal, show for the same period fifty shocks, a number of which were severe. (P. 11, S. Rep. 783, part 2, 57th Cong., 1st session, May 31, 1902.)

In another part of their report the committee said:

With the dreadful lessons of Martinique and St. Vincent fresh in our minds, we should be utterly inexcusable if we deliberately selected a route for an Isthmian canal in a region so volcanic and dangerous, when a route is open to us which is exposed to none of these dangers and is in every other respect more advantageous.

And they quote Professor Heilprin, an authority on the subject, in part, as follows:

It has, however, been known for a full quarter of a century that the main Andes do not traverse the Isthmus of Panama, and that there are no active or recently decayed volcanoes in any part of the Isthmus. So far, however, as danger from direct volcanic contacts is concerned, the Panama route is exempt. (Pp. 22–23.)

And further:

This district represents the most stable portion of Central America. No volcanic eruptions have occurred there since the end of the Miocene epoch, and there are no active volcanoes between Chiriqui and Tolima, a distance of about four hundred miles. Such earthquakes as have occurred are chiefly those proceeding from the disturbed districts on either hand, with intensity much diminished by the distance traversed. The canal lies in a sort of dead angle of comparative safety.

The report continues:

The situation being, then, that the danger from volcanoes at Panama is nothing, and that from earthquakes practically nothing, while at Nicaragua the canal would be situated in one of the most dangerous regions of the world from both these causes, the question should be considered settled.

This was the opinion of the committee of 1902; it was emphatic and plain in its language; it had considered expert views and the available data. It had before it the full report of the Nicaragua Canal Commission, printed under date of May 15th of the same year, Chapter VII of which considers the subject at much greater length than has been done since that time and with a full knowledge of the facts and free from bias or prejudice. With the then recent occurrence at Mount Pelée in mind, and with a full understanding of the liability of the Isthmus to seismic shocks of minor importance, the committee emphatically indorsed the lock-canal project at Panama.

Much can be said with regard to this matter, and it is one which should, and no doubt will, receive the most careful consideration of the engineers in charge of the work. Seismic disturbances have occurred in all parts of the world, and they have occurred at Panama. Where they are not directly of volcanic origin they appear to be the result of subsidence or contraction of the earth's crust, and they have occurred and caused serious destruction far from centers of volcanic activity, among other places, at Lisbon, Portugal, and at Charleston, S.C. Some sections of the earth, as for illustration Japan and the Philippines, are no doubt more subject to these movements than others, and sections subject to such movements at one period of time may be exempt for many years if not forever thereafter.

The fearful earthquake which affected Charleston, S.C., in 1886 had no corresponding precedent in that section, nor has it been followed by a similar disturbance. Regardless of the terrible experience of 1886, the government has now in course of construction at Charleston a navy-yard, and a great dry-dock, costing many millions of dollars, which will be operated by locks or gates, and, I presume, the question of earthquakes or earth movements has not been raised in any of the reports which have been made regarding this undertaking. Earthquakes formerly were quite frequent in New

England, and they extended to New York during the early years of our history, and for a time Boston and Newbury, Mass., Deerfield, N.H., and particularly East Haddam, Conn., were the centers of seismic activity, which by inference might be used as an argument against our navy-yards at Portsmouth, N.H., and Charlestown, Mass., our torpedo station at Newport, or the fortifications at Willets Point. The earthquake which destroyed Lisbon in 1755 might with equal propriety be used as an argument against the building of the extensive docks and fortifications at Gibraltar, but no one, I think, has ever questioned the solidity of the Rock.

Seismology is a very complex branch of geologic inquiry and it is a subject regarding which very little of determining value is known. Theories have been advanced that under certain geological conditions earth movements would be comparatively infrequent, if not impossible. Whether such conditions exist at Panama would have to be determined by the investigations of qualified experts. It would seem, however, from such data as are available, that the local conditions are decidedly favorable to a comparative immunity of this region from serious seismic shocks, at least such as would do great and general damage. Nor can it be argued that the locks and dams would be exposed to special risk. The earthquake of 1882 did more or less damage, but the reports are of a very fragmentary character. Newspaper reports in matters of this kind have very small value. Injury was done to the railway, but not of very serious consequence.

If the risk exists, it would affect equally a sea-level canal, in that it would threaten the tidal lock, the dam at Gamboa, and the excavation through Culebra cut. Very little is known regarding earthquake motions, and there are very few seismic elements which are really calculable in conformity to a mathematical theory of probability. It is a subject which has not received the attention in this country of which it is deserving, but enough of seismic motion is known to warrant the conclusion that the Senate committee of 1902 was, in all human probability, entirely correct when it made light of the danger of the probability of seismic shocks at Panama.

In fine, the earthquake argument has little or no force against a lock-canal project, and it has never received serious consideration as such or been used in arguments against a lock canal until the recent

San Francisco disaster brought the subject prominently before the public. It is a danger as remote as a possible destruction of the proposed terminal plants at Colon and Panama by flood waves equal in magnitude to the one which destroyed Galveston in 1900, but such dangers are inherent in all human undertakings. They must be taken as a matter of chance and remote possibility, which for all present purposes may be left out of account, except that the subject should receive the due consideration of the engineers and perhaps be made a matter of special and comprehensive inquiry by the Geological Survey. In any serious consideration of the facts for or against a lock canal, I am confident that the earthquake risk may safely be ignored. The comprehensive report of the minority members of the Senate Committee on Interoceanic Affairs is a sufficient and conclusive answer to all the important points which are in controversy, and it remains for Congress to cut the "Gordian knot" and put an end to an interminable discussion of much solid and substantial conviction on the one hand and of a vast amount of opinion and guesswork on the other hand. All of the evidence, all of the supplementary expert testimony which may be obtained upon the merits of the two propositions, will not change the position of those who rest their conclusions upon American experience and upon the judgment of American engineers, and who favor a lock canal. While there is no doubt that such a canal can be constructed and can be made a practicable waterway, there is a very serious question whether a sea-level canal can be constructed and made a safe and practicable waterway, at least within the limits of the estimated amount of cost and within the estimated time.

The view which I have tried to impress upon the Senate is nothing more nor less than a business view of what is, for all practical purposes, only a business proposition. If a lock canal can be built, useful for all purposes, at half the cost and within half the time of a sea-level canal, then I can come to no other conclusion than that a lock canal would be decidedly to our political and commercial advantage. A decision, however, should be arrived at. The canal project has reached a stage where the final plan or type must be determined, and it is the duty of Congress to act and to fix, for once and for all time, the type of canal, with the same courage and freedom from prejudice or bias as was the case in the decision which finally fixed the route by way of Panama.

Any amount of additional testimony and expert opinion will only add to the confusion and tend to produce a more hopeless state of affairs. Let Congress fix the type in broad outlines and leave it to responsible engineers in actual charge to solve problems in detail, and to adapt themselves to local conditions and to new problems which in the course of construction are certain to arise. Let us take counsel of the past, most of all from the experience gained in the construction of the Suez Canal, an engineering and commercial success which challenges the admiration of the world. We know how near it came to utter defeat by the conflict of opinion, by the intrigue of conniving and jealous powers, and last, but not least, by the ill-founded apprehensions and fears of those who were searching the vast domain of conjecture and remote possibilities for arguments to cause a temporary delay or ultimate abandonment.

It is not difficult to secure the opinion of eminent authority for or against any project when the facts themselves are in dispute, and when the objects and aims are not well defined. The great Lord Palmerston, the most bitter opponent of the Suez Canal scheme, in want of a more convincing argument, seriously claimed that France would send soldiers disguised as workmen to the Isthmus of Suez, later to take possession of Egypt and make it a French colony. By one method or another Palmerston tried to defeat the scheme in its beginning and to bring it to disaster during the period of construction. It is a far from creditable story. History always more or less repeats itself, whether it be in politics or engineering enterprise, but in few affairs are there more convincing parallels than in the canal projects of Panama and Suez. Lord Palmerston and Sir Henry Bulwer, then the ambassador at Constantinople, did all in their power to destroy public confidence in the enterprise, and they were completely successful in preventing English investments in the stock of the canal. [2]

It was the same Sir Henry Bulwer who, in 1850, succeeded by questionable diplomatic methods in foisting upon the American people a treaty which was contrary to their best interests and which for half a century was a hindrance and barrier to an American Isthmian canal. We owe it chiefly to the masterly and straightforward statesmanship of the late John Hay that this obstacle to our progress was disposed of to the entire satisfaction of both nations. I refer to these matters, which are facts of history, only to point out how an interminable discussion of matters of detail is certain to delay and do great injury to projects which should only receive Congressional consideration in broad outlines and upon fundamental principles. If we are to enter into a discussion of engineering conflicts, if we are to deliberate upon mere matters of structural detail, then an entire session of Congress will not suffice to solve all the problems which will arise in connection with that enterprise in the course of time. I draw attention to the Suez experience solely to point out the error of taking into serious account minor and far-fetched objections which assume an undue magnitude in the public mind when they are presented in lurid colors of impending disasters to a national enterprise of vast extent and importance.

So eminent an engineer as Mr. Robert Stephenson by his expert opinion deluded the British people into the belief that the Suez Canal would not be practicable; that, even if completed, it would be nothing but a stagnant ditch. Said Palmerston to De Lesseps:

All the engineers of Europe might say what they pleased, he knew more than they did, and his opinion would never change one iota, and he would oppose the work to the end.

Stephenson confirmed this view and held that the canal would never be completed except at an enormous expense, too great to warrant any expectation of return—a judgment both ill advised and erroneous as was clearly proved by subsequent events. I need only say that the Suez Canal is to-day an extremely profitable waterway, and that although the work was commenced and brought to completion without a single English shilling, through French enterprise and upon the judgment of French engineers, it was only a comparatively few years later when, as a matter of necessity and logical sequence, the controlling interest in the canal was purchased by the English government, which has since made of that waterway the most extensive use for purposes of peace and of war.

These are the facts of history, and they are not disputed. Shall history repeat itself? Shall we delay or miscarry in our efforts to complete a canal across the Isthmus of Panama upon similar pretensions of assumed dangers and possibilities of disaster, all more or less the result of engineering guesswork? Shall we take fright at the talk about the mischief-maker with his stick of dynamite, bent upon the destruction of the locks and the vital parts of the machinery, when history has its parallel during the Suez Canal agitation in "the Arab shepherd, who, flushed with the opportunity for mischief and with a few strokes of a pickax, could empty the canal in a few minutes"? Shall we be swayed by foolish fears and apprehensions of earthquakes or tidal waves, and waste millions of money and years of time upon a pure conjecture, a pure theory deduced from fragmentary facts? Again the facts of canal history furnish the parallel of Stephenson and other engineers, who successfully frightened English investors out of the Suez enterprise by the statement that the canal would soon fill up with the moving sands of the desert, that one of the lakes through which the canal would pass would soon fill up with salt, that navigation of the Red Sea would be too dangerous and difficult, that ships would fear to approach Port Said because of dangerous seas, and, finally, that in any event it would be impossible to keep the passage open to the Mediterranean.

It was this kind of guesswork and conjecture which was advanced as an argument by engineers of eminence and sustained by one of the foremost statesmen of the century. How absurd it all seems now in the sunlight of history! The Panama Canal is a business enterprise, even if carried on by the nation, and with a thorough knowledge of the general facts and principles we require no more expert evidence, so called, nor additional volumes of engineering testimony. The nation is committed to the construction of a canal. The enterprise is one of imperative necessity to commerce, navigation, and national defense, and any further discussion, any needless waste of time and money, is little short of indifference to the national interests and objects which are at stake.

Of objections to either plan there is no end, and there will be no end as long as the subject remains open for discussion. To answer such objections in detail, to search the records for proof in support of one theory or another, is a mere waste of time which can lead to no possible useful result. Among others, for illustration, there has been placed before us a letter from the chief engineer of the Manchester Ship Canal, who is emphatically in favor of a sea-level waterway. It would have been much more interesting and much more valuable to the members of Congress to have received from Mr. Hunter a statement as to why he should have changed his opinions; or why, in 1898, he should have signed the unanimous report of the technical commission in favor of a lock canal, while now he so emphatically sustains those who favor the sea-level project. It is not going too far to say, appealing to the facts of history, that Mr. Hunter may be seriously in error in this matter and may have drawn upon his imagination rather than upon his engineering experience, the same as Mr. Robert Stephenson was in serious error in his bitter opposition to the canal enterprise at Suez.

Mr. Hunter, in his letter, argues, among other points, that the lifts of the proposed locks would be without precedent. Without precedent? Why, of course, they would be without precedent. Is not practically every large American engineering enterprise without precedent? Was not the Erie Canal, completed in 1825, without precedent? Were not the first steamboat and the first locomotive without precedent? Were not the Hoosac Tunnel and the Brooklyn Bridge feats of American engineering enterprise without precedent?

Without precedent is the great barge canal which the State of New York is about to build, which will mean a complete reconstruction of the existing waterway which connects the ocean with the Great Lakes. [3]

All this is without precedent. But it is American. It is progress, and takes the necessary risk to leave the world better, at least in a material way, than we found it. In the proposed deep waterway, which is certain some day to be built to connect the uttermost ends of the Great Lakes with tide-water on the Atlantic, able and competent engineers of the largest experience have designed locks with a lift of 52 feet. [4] That will be without precedent. On the Oswego Canal, proposed as a part of the new barge canal of the State of New York, there will be six locks, two of which will each have a lift of 28 feet, [5] and that will be without precedent, but neither dangerous nor detrimental to navigation interests.

Need I further appeal to the facts of past canal history? Is it necessary to recite one of the best known and most honorable chapters in the history of inland waterways—I mean the problems and difficulties inherent in the great project of constructing the canal of Languedoc, or "Canal du Midi," which forms a water communication between the Mediterranean and the Garonne and between the Garonne and the Atlantic Ocean, one of the best known canals in France and in the world? Need I refer to that pathetic story of its chief engineer, Riquet, one of the greatest of French patriots, who, in his abiding faith in this great engineering feat, stood practically alone? Need I recall that he met with scant assistance from the government, with the most strenuous opposition from his countrymen; that he was treated even as a madman and that he died of a broken heart before the great work was finished?

That canal stands to-day as an engineering masterwork and as a most suggestive illustration of man's ingenuity and power to overcome apparently insuperable natural obstacles. It has been in existence and successful operation, I think, since 1681. For a sixth part of its distance it is carried over mountains deeply excavated. It has, I think, ninety-nine locks and viaducts, and as one of its most wonderful features it has an octuple lock, or eight locks in flight, like a ladder from the top of a cliff to the valley below. If in 1681 a French engineer had the ability and the daring to conceive and construct an octuple lock, will any one maintain that more than two hundred years later, with all the enormous advance in engineering, with a better knowledge of hydraulics and a more perfect method of transportation and handling of materials—will any one maintain that we are not to-day competent to construct successfully a lock canal such as is proposed to be built at Panama upon the judgment of American engineers?

Mr. President, the overshadowing importance of the subject has led me to extend my remarks far beyond my original intention. I express my strong convictions in favor of a lock canal and of the necessity for an early and specific declaration of Congress regarding the final plan or type of canal which the nation wants to have built at Panama. I am confident that it lies entirely within our power and means to build either type of waterway; that our engineering skill can successfully solve the technical problems involved in either the lock or the sea-level plan; but there is one all-important factor which controls, and which, in my opinion, should have more weight than any other, and that is the element of time. If I could advance no other reasons, if I knew of no better argument in favor of a lock canal, my convictions would sustain the project which can be completed within a measurable distance of years and for the benefit and to the advantage of the present generation. Time flies, and the years pass rapidly. Shall this project languish and linger and become the spoil of political controversy and a subject of political attack? Can we conceive of anything more likely to prove disastrous to the canal project than political strife, which proved the undoing of the French canal enterprise at Panama?

Shall the success of this great project be imperiled by the possible changes in the fortunes of parties? Shall we incur the risk that changes in economic conditions, hard times, or panic and industrial depressions may bring about? Time flies, and in the progress of industry and commerce, in international competition and the growth of modern nations, no factor is of more supreme importance than the years, with new opportunities for political and commercial development. Shall we, then, neglect our chances? Shall we fail to make the most of this the greatest opportunity for the extension of our commerce and navigation into the most distant seas which will ever come to us in our history, because of the demands of idealists, who, with theoretical notions of the ultimately desirable, would deprive the nation and the world of what is necessary and indispensable to those who are living now?

Vast commercial and political consequences will follow the opening of the transisthmian waterway. In the annals of commerce and navigation it is not conceivable that there will ever be a greater event or one fraught with more momentous consequences than uninterrupted navigation between the Atlantic and the Pacific. Little enough can we comprehend or anticipate what the far-distant future will bring forth, but this much we know—that it is our duty to solve the problems of *to-day* and not to indulge in dreams and fancies in a vain effort to solve the problems of a far-distant future.

But *money* also counts. Can we defend an expenditure of an additional $100,000,000 or more for objects so remote, and upon a basis of theory and fact so slender and so open to question, when a plan and a project feasible and practicable is before us which will meet all of our needs and the needs of generations to come? Shall we disregard in the building of this canal every principle of a sound national economy and commit ourselves to an enormous waste of funds and to the imposition of needless burdens upon the taxpayers of this nation and upon the commerce of the world? At least $2,000,000 more per annum will be required in additional interest charges, at least $100,000,000 more will be necessary as an original investment. Do we fully realize what that amount of money would do if applied to other national purposes and projects?

I want to place on record my convictions and the reasons governing my vote in favor of the minority report for a lock canal across the Isthmus at Panama. I entered upon an investigation of the subject without prejudice or bias and have examined the facts as they have been presented and as they are a matter of record and of history. I have heard or read with care the evidence as it has been presented by the Board of Consulting Engineers and the vast amount of oral testimony before the Senate Committee on Interoceanic Affairs. I am confident that the minority judgment is the better and that it can be more relied upon, because it is strictly in conformity with the entire history of the Isthmian canal project. I am confident that the objections which have been raised against the lock plan are an undue exaggeration of difficulties such as are inherent in every great engineering project, and which, I have not the slightest doubt, will be successfully solved by American engineers, in the light of American experience, exactly as similar difficulties have been solved in many other enterprises of great magnitude.

I am not impressed with the reasons and arguments advanced by those who favor the sea-level project, for they do not appeal to me as being sound, and in some instances they come perilously near to being engineering guesswork characteristic of the earlier enterprises of De Lesseps. I cannot but think that bias and prejudice are largely responsible for the judgment of foreign engineers so pronounced in favor of a sea-level project. Furthermore, I am entirely convinced that the judgment and experience of American engineers in favor of a lock canal may be relied upon with entire confidence, and that such an enterprise will be brought to a successful termination. I believe that in a national undertaking of this kind, fraught with the gravest possible political and commercial consequences, only the judgment of our own people should govern, for the protection of our own interests, which are primarily at stake. I also prefer to accept the view and convictions of the members of the Isthmian Commission, and of its chief engineer, a man of extraordinary ability and large experience.

It is a subject upon which opinions will differ and upon which honest convictions may be widely at variance, but in a question of such surpassing importance to the nation, I, for one, shall side with those who take the American point of view, place their reliance upon American experience, and show their faith in American engineers.

Founded by JOHN F. DRYDEN
Pioneer of Industrial Insurance in America

THE PRUDENTIAL INSURANCE COMPANY OF AMERICA

Incorporated under the laws of the State of New Jersey

FORREST F. DRYDEN, *President*
HOME OFFICE, NEWARK, NEW JERSEY

FOOTNOTES:

[1] Report of the New Panama Canal Company of France; Senate Document 188, 56th Congress, 1st session, February 20, 1900.

[2] The Maritime Canal of Suez, from its inauguration, November 17, 1869, to the year 1884, by Prof. J.E. Nourse, U.S.N., Washington, 1884 (Senate Document 198, 48th Congress, 1st session).

[3] For a history of American canal building enterprises see History of New York Canals, ch. 5.

[4] Report of the Board of Engineers on Deep Waterways, H. of R., Doc. No. 149, 56th Congress. 2d session, Atlas.

[5] History of New York Canals, Appendix L. Annual Report of the State Engineer and Surveyor. Vol. II, Albany, N.Y., 1905.

www.ingramcontent.com/pod-product-compliance
Lightning Source LLC
Chambersburg PA
CBHW030507220526
45464CB00006B/2699